CITY MACHINES
STREET SWEEPERS

Connor Dayton

PowerKiDS
press.

New York

Published in 2012 by The Rosen Publishing Group, Inc.
29 East 21st Street, New York, NY 10010

First Edition

Editor: Jennifer Way
Book Design: Ashley Drago

Photo Credits: Cover, pp. 4–5, 10–11, 14, 18–19, 23, 24 (top left) Shutterstock.com; pp. 6, 9, 24 (bottom left) iStockphoto/Thinkstock; p. 13 © www.iStockphoto.com/David H. Lewis; p. 17 © www.iStockphoto.com/Jim Pruitt; p. 20 © www.iStockphoto.com/flashgun; p. 24 (top right) PhotoDisc; p. 24 (bottom right) © www.iStockphoto.com/Mike Clarke.

Library of Congress Cataloging-in-Publication Data

Dayton, Connor.
 Street sweepers / by Connor Dayton. — 1st ed.
     p. cm. — (City machines)
 Includes index.
 ISBN 978-1-4488-4961-1 (library binding) — ISBN 978-1-4488-5072-3 (pbk.) —
 ISBN 978-1-4488-5073-0 (6-pack)
 1. Street cleaning—Equipment and supplies—Juvenile literature. I. Title.
 TD860.D39 2012
 628.4'6—dc22
                                                    2010051384

Manufactured in the United States of America

CPSIA Compliance Information: Batch #WS11PK: For Further Information contact Rosen Publishing, New York, New York at 1-800-237-9932

CONTENTS

Street sweepers clean city streets.

4

Street sweepers are trucks. They have special cleaning **brushes**.

Jets spray water. Then the brushes wash the street.

Charles Brooks invented the street sweeper. He was an African American from Newark, New Jersey.

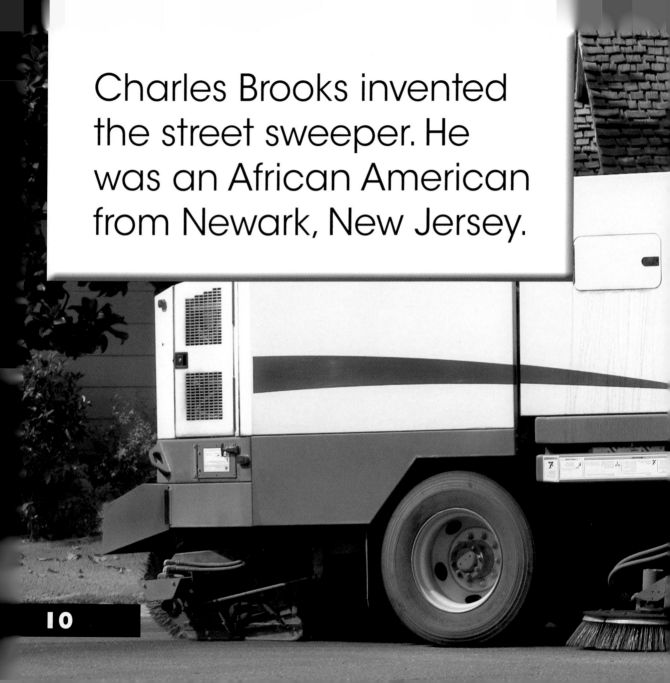

The street sweeper picks up dirt and **trash**. They go into a part called the hopper.

Small street sweepers have three wheels. They are used in small spaces.

Small street sweepers can clean sidewalks, too.

17

Big street sweepers clean big streets. They even clean **highways**!

Most cities have street sweepers. Boise, Idaho, was the first U.S. city to have one.

Street sweepers are important machines. They keep cities clean!

WORDS TO KNOW

brushes

highway

jets

trash

INDEX

WEB SITES

Due to the changing nature of Internet links, PowerKids Press has developed an online list of Web sites related to the subject of this book. This site is updated regularly. Please use this link to access the list:
www.powerkidslinks.com/city/street/